Basics of Addition and Subtraction

Preface

Elementary school math! Can we learn in a more interesting way that is more meaningful to us?

Many students think math is boring and difficult. It is not that unusual for us to meet people who gave up on math. Given this reality, this book was made with a teacher's enthusiasm in teaching students elementary math in a more interesting and meaningful way.

This book contains at least three characteristics. First, based on Korean math and education procedures, materials are compared to Chinese, American, and Japanese math curriculum and analyzed. The vital elements of elementary math were extracted. Therefore, you can learn general math content through this book series. Second, we developed the manuscript by considering how students acquire math content and apply the steps and methods to do so, even examining their behavior with math. Third, considering students' interests and curiosity those factors extracted are reflected in an interesting fantasy story.

Researchers and writers of this manuscript took a meaningful journey during its preparation. We anticipate that all students who are reading this book will go on a journey with Ryan and Aris to learn math and realize that math is fun and meaningful in our world.

Korea National University of Education (Math Department) Professor
Bang Jung Sook

Preface

I will become a revolutionist who will shake the stereotype of studying being boring!

While I was in school, I didn't know the reason why I had to learn math. Memorizing the formulas and solving questions repeatedly, I loathed math at my young age. After becoming an adult, I was surprised that all mathematical principals are blended into our life. In reality, math has a close relationship with us, yet our children do not know such facts and don't see the connection. Therefore, they lose interest and even get stressed on solving questions and the scores they get. Eventually, they lose confidence in math.

This led to the forming of a team of math teachers, education specialists, and professional writers to make *Ryan's Math Adventure*, offering students a way to gain confidence and enjoy math.

We analyzed the education procedures of Korea, China, America, and Japan and made a conceptual and global curriculum. And through the deployment of a fantasy adventure, we solve mathematical questions and connect the abstract ideas of math and our reality. Children forget that they are learning math and enjoy this comic book style, naturally understanding the principles of math within. They complete study goals by using a workbook to establish concepts and solve interesting questions. Also, through a fantasy world with various types of beings and cultures and a European style of illustration, without violence, that we consider necessary for children to have a positive mindset and behavior toward math and a multi-language system, I hope our children grow up to become global human resources for our bright future.

WeDu Communications CEO
Lee Kyu Ha

Composition & Features
Integrated Global Education Course

Through comparative analysis of math curriculum of Korea, China, Japan, and America, the Integrated Global Education Course is designed for a new math "edutainment" book. With contents intended to reflect world trends and to provide the objective rationale, the Integrated Global Education Course can target a broad range of readers.

Mathematical Contents, Elements, and Composition

❶ Based on the math learning achievement standards presented in the curriculum of China, Japan, and America, we have developed concrete mathematics contents which elementary students should learn.

❷ By analyzing the other three countries' curriculum based on Korea's math learning achievement standards and textbook contents, we have identified the most commonly emphasized math contents, which are reflected in the book.

❸ Other details contained in the other three countries' curriculum but not in Korea's are also identified; noteworthy differences are captured in the book for a truly in-depth aggregate composition.

Goals of Integrated Global Education Course

❶ In the course of observing, analyzing, organizing, and expressing various phenomena around us, learners will develop abilities to discover mathematical concepts, functions, principles, and laws inherent in those phenomena and to understand the correlations between them.

❷ The capability to mathematically think, express, and communicate will be fostered; learners will be able to solve various issues derived from different phenomena in creative and reasonable ways.

❸ Learners will be able to understand values of mathematics; they will look upon math as a joy; and a good-natured personality will be nurtured.

Most children regard mathematics as a boring subject where memorization and calculation are heavily involved. Unlike their expectations, *Ryan's Math Adventure* provides fun math learning through storytelling techniques which help learners in learning math concepts and principles while reading comics. Learners will be able to adopt a flexible, creative, mathematical way of thinking while exploring a fantasy land.

1. By closely correlating fun fantasy stories and math concepts, this book empowers children to understand mathematical principles in a fun and natural way.

2. Through daily life events, children will learn how deeply math is related to our everyday life.

3. The main characters of the book deal with not only the direct mathematic area but also mathematical competence emphasized in elementary school mathematics curriculum including problem solving, reasoning, and mathematical communication, and more. Therefore, as each character grows in math capability and good manners, learners will achieve the same growth.

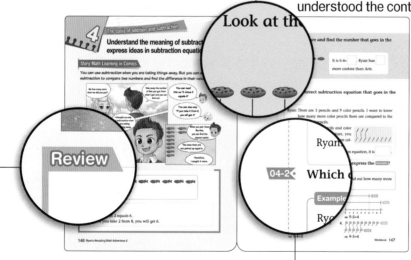

Comprehensive exercises will be provided to see how well children understood the contents.

An explanation matching to each curriculum is provided to help learners gain a full understanding of math concepts.

Extended exercises will be provided for children to have an in-depth understanding and to let them get used to various types of math practice.

Introduction of Characters

Ryan
(10 years old)

A normal student
from our world.
Loves fantasy novels, games,
cartoons,
and movies, but hates math.
Usable item: Yo-Yo

Aris
(7 years old)

A curious princess
from Tanare Palace
that lives in this
fantasy world.
King Adel,
who is her father,
gives her the *Star of Wisdom* as a gift.
Usable item: Magical Wand

Phillip
(10 years old)

A noble man
from the fantasy world.
Excellent with numbers
and operations.
Like his peers,
Phillip is excellent with magic
and swordsmanship.
Usable item: Sword

Numi
(10 years old)

A member of the Elf Clan,
loves adventures and
is very active. Has very good
knowledge of
diagrams and uses
diagram magic.
Usable item: Bow

Pabel
(10 years old)

A member of the Dwarf Clan which had the knowledge of measurement but, after the magical curse of Pesia, lost all knowledge of math.

Usable item: Axe, Hammer

Gilly
(13 years old)

A member of the Floa Clan who can transform into a tree. Knows basic math and learned magic.

Usable item: Asian Guitar

Walter
(33 years old)

Royal general of Tanare Palace. Excellent with math and magic. Good with machines and builds a robot for Aris.

Namute

A smart robot created by Walter to protect Aris. But after interacting with Ryan, Namute becomes a weird robot.

Special Skill: Transform into a motorcycle

Villains of this book

Pesia

A person full of desire to become the only king and to rule all. With the *Staff of Chaos* in his possession, he erases nearly all knowledge of math from the world. He threatens the journey of Ryan and Aris, looking to get Aris's necklace, the *Star of Wisdom*, at any cost.

Usable Item: *Staff of Chaos*

Sirocco

A faithful servant to Pesia, he comes from the same town as Walter. However, Sirocco is always number 2, due to Walter, so he has great anger and jealousy toward him.

Dagan

He is close to Lord Dior of Onix but his true identity is as servant of Pesia. He works as an informant and passes the secret of the *Book of Light* to Pesia.

Nurimas

The only blood relative of Pesia - a nephew. Since he was young, he has been raised by Pesia so that now he is cold and decisive, and obeys Pesia no matter what.

Past story

Ryan is a boy from our world who hates math. He teleported to a fantasy world with a book he found at the museum. There, the peaceful kingdom of Tanare has been overtaken by a villain named Pesia. Pesia uses his staff of Chaos to try to erase all knowledge of math from the world. A world without math is a world in great disorder. Our heroes, Walter and Aris, barely make it out of the castle and away from Pesia. Ryan meets Walter and Aris and together they start a journey. They find the Book of Light, just like the old prophecy said. Ryan, Aris, and Walter arrived at Revna, Walter's hometown. They meet Phillip, who is the grandson of the Lord of Onix. The town called Revna is a mess since no one knows math any longer. Ryan, Aris, and Walter go to the hospital and the robot factory to solve some problems. Walter makes a robot for Aris. Aris names the robot Namute. But just as the Stone of Life was being inserted into the robot, it accidentally got hit by Ryan's yo-yo. Now Namute can think and act independently and it is somewhat unorthodox.

contents

1. Reunion with Gilly

① Understanding Addition (1) ··· 12
② Understanding Addition (2) ··· 27

2. Dangerous larva monsters

③ Understanding Subtraction (1) ··· 36
④ Understanding Subtraction (2) ··· 44
⑤ Adding and Subtracting 0 ··· 49

3. Suspicious Guy

⑥ Understanding the Commutative Law of Addition ··· 70
⑦ Number Pairs of 10 (1) ··· 76
⑧ Number Pairs of 10 (2) ··· 85
⑨ Adding (Tens)+(Ones) ··· 90
⑩ Adding Three Numbers (The Associative Law) ··· 97
⑪ Calculating (Ones)+(Ones)=(Tens) ··· 104

4. Yoyo and magic wand

⑫ Calculating (Tens)−(Ones) ··· 112
⑬ Subtracting Three Numbers ··· 119
⑭ Adding and Subtracting Three Numbers ··· 125
⑮ Calculating (Tens)−(Ones)=(Ones) ··· 131

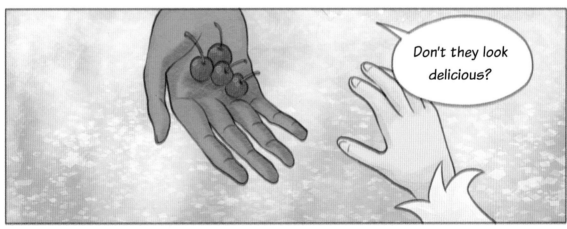

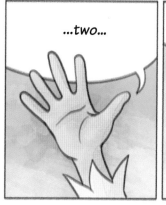
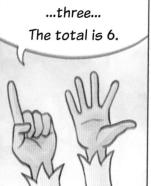

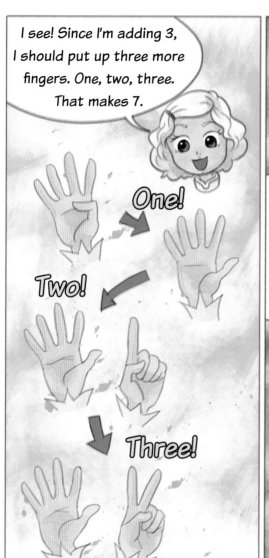

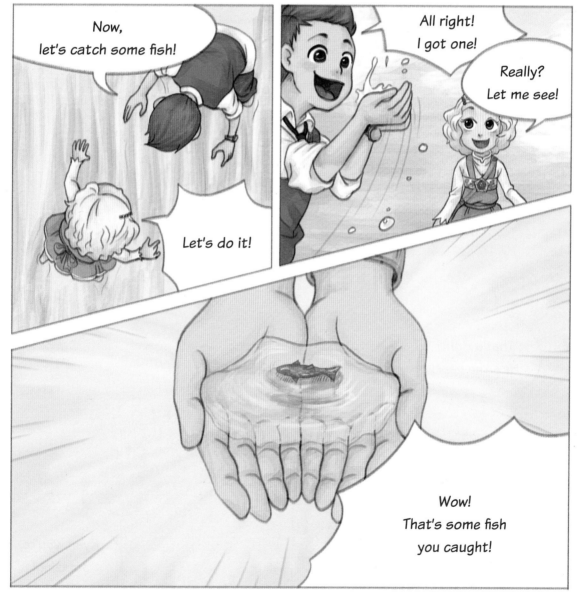

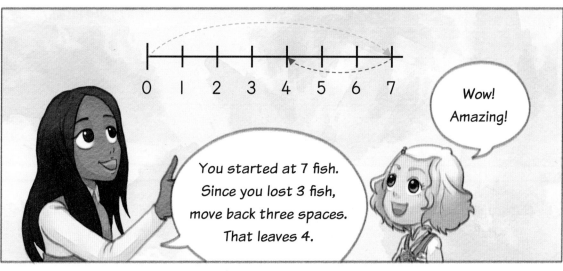

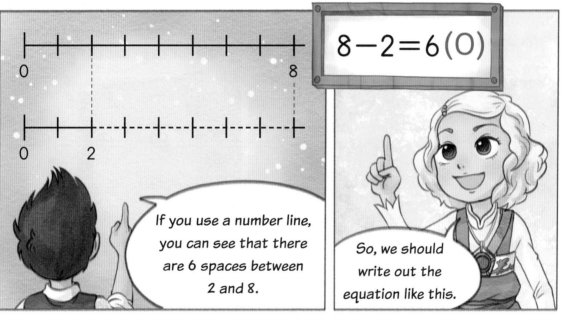

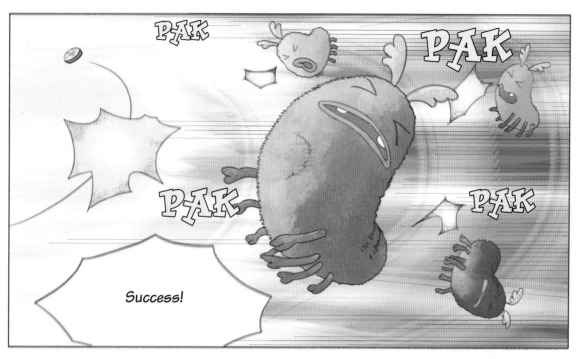

2. Dangerous larva monsters

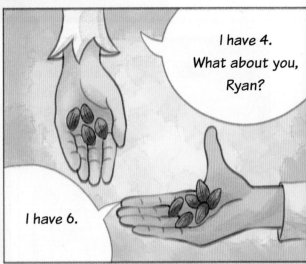

How many books did you bring?

3. You?

7 for me.

I brought 6 myself.

How many books did we bring altogether?

You can express this as an addition equation.

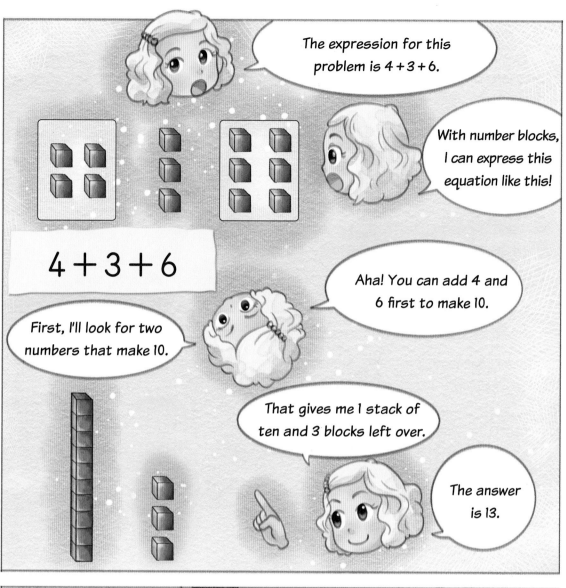

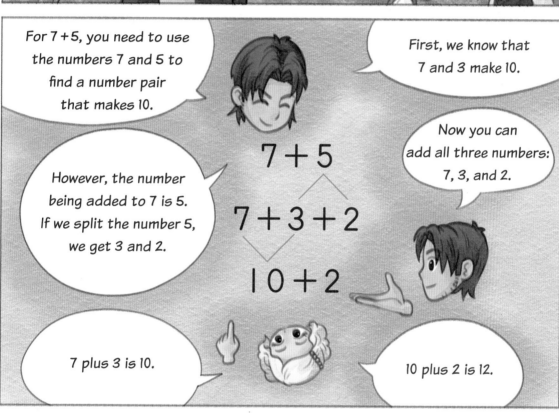

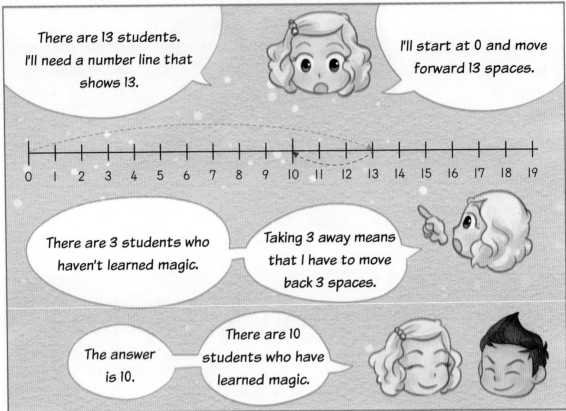

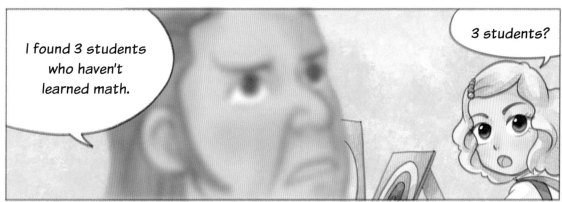

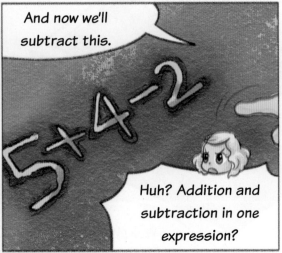

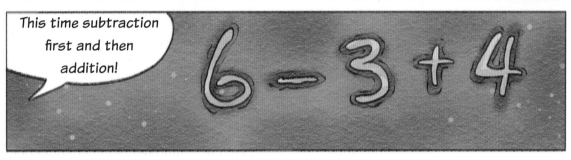

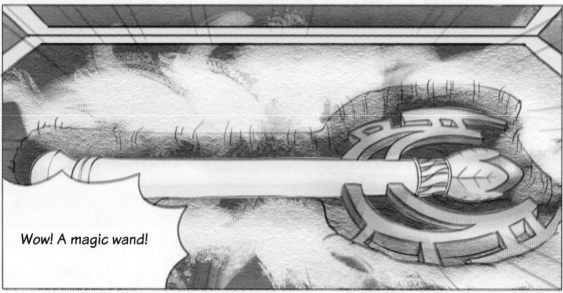

131

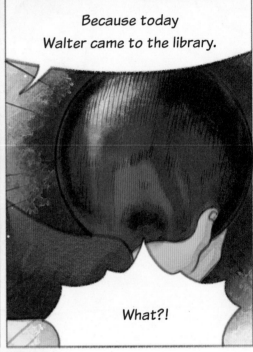

Basics of Addition and Subtraction

Understanding Addition (1)

Comics for Learning Math

Aris had 3 cherries. Gilly brought 4 more cherries. Put a "+" sign to show that two numbers are being added together.

☆ Addition means you are putting together, or combining, a number and another number.

Review

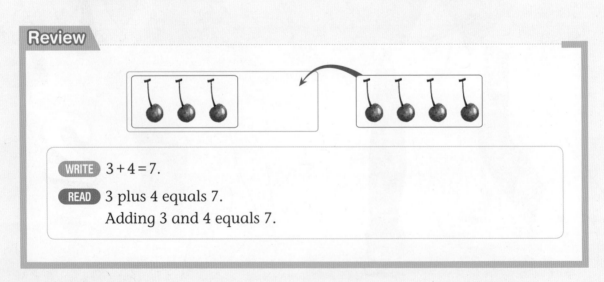

WRITE $3 + 4 = 7$.

READ 3 plus 4 equals 7.
Adding 3 and 4 equals 7.

Warm Up 01 Look at the picture and choose the correct answer.

$3 + 1 = 4$

3 plus 1

(equals , does not equal) 4.

01-1 Look at the picture. Which jar of beads best fits A?

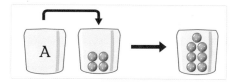

$4 + 3 = 7$

a.   b.   c.   d.   e.

01-2 Which of the following explanations is <u>NOT</u> correct according to the number line?

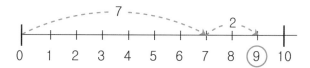

a. $7 + 2 = 9$

b. 7 plus 2 equals 9.

c. When you move 7 spaces forward from 2, you will stop at 10.

d. Adding 7 and 2, you will get 9.

2 Basics of Addition and Subtraction

Understanding Addition (2)

Comics for Learning Math

An " = " sign means that the values on either side equal each other.

When the values of both sides are equal, you can use an equal sign.

Review

| 🍍🍍🍍🍍 | 🍍🍍🍍 |

WRITE 4 + 3 = 7

READ 4 plus 3 equals 7.
Adding 4 and 3 equals 7.

Warm Up 02 Which addition expression matches the picture?

a. 3 + 2   b. 4 + 2   c. 2 + 4   d. 5 + 1

02-1 Which of the following pictures does NOT match the addition equation next to it?

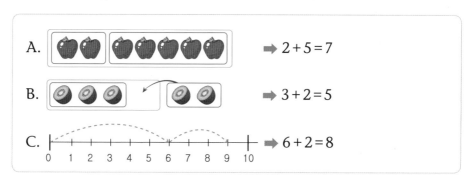

A. ➡ 2 + 5 = 7

B. ➡ 3 + 2 = 5

C. ➡ 6 + 2 = 8

02-2 Read the following sentences. Which does NOT match the picture?

a. In the lion cage, there are 2 lionesses and 5 lions. There are 7 lions in total.

b. There are 2 children eating ice cream and 3 children standing in line to buy ice cream. There are 5 children altogether.

c. There are 6 sparrows sitting in the trees and 2 sparrows flying. There are 9 sparrows in total.

3. Basics of Addition and Subtraction

Understanding Subtraction (1)

Comics for Learning Math

You can use a " – " sign when you are taking a number away from another number.

You can also say "3 subtracted from 7 is 4."

You read it as "7 minus 3 equals 4."

You can use subtraction to find out how many fish remain.

Review

WRITE $7 - 3 = 4$

READ 7 minus 3 equals 4.
3 subtracted from 7 is 4.

Warm Up 03 There were 5 melons. 3 of them were eaten. Choose the subtraction expression that equals the remaining number of melons.

a. 2+3   b. 3+2   c. 5−3   d. 5−2   e. 3−2

03-1 Which of the subtraction equations is correct according to the number line?

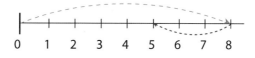

a. 8−1=7   b. 8−5=3   c. 8−3=5
d. 5−3=2   e. 8−0=8

03-2 Read the story below. Choose the answer that matches the story.

> **Story**
> There were 6 cookies. You gave 2 of them to a friend.

a.  , 6−2=4

b. , 6−4=2

c. , 6−2=4

d. , 6−4=2

Basics of Addition and Subtraction

Understanding Subtraction (2)

Comics for Learning Math

You can use subtraction when you are taking away a number from an original number. You can also use subtraction to compare the difference in values between two numbers.

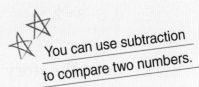

You can use subtraction to compare two numbers.

Review

WRITE $8 - 2 = 6$

READ 8 minus 2 equals 6.
2 subtracted from 8 is 6.

Warm Up 04 — Look at the picture and write the number that fits in both boxes.

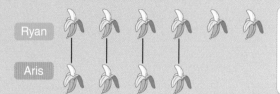

It is 6 − 4 = ☐. Ryan has ☐ more bananas than Aris.

04-1 Write the correct subtraction equation in the box.

Ryan: There are 3 red pencils and 9 blue pencils. How many more blue pencils do I have?

Aris: If you pair the red pencils and blue pencils like in the picture, you can see that you have 6 more blue pencils.

Ryan: If you write this as a subtraction equation, it is ☐.

04-2 Which of the following pictures matches the Story?

Story

Ryan has 9 oranges, and Aris has 5. How many more oranges does Ryan have?

a.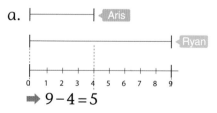
➡ 9 − 4 = 5

b.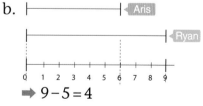
➡ 9 − 5 = 4

c.
➡ 9 − 4 = 5

d.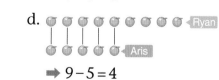
➡ 9 − 5 = 4

5 Basics of Addition and Subtraction

Adding and Subtracting 0

Comics for Learning Math

If you add 0 to a number, the value does not change.

★ If you add or subtract 0, the original number does not change.

Review

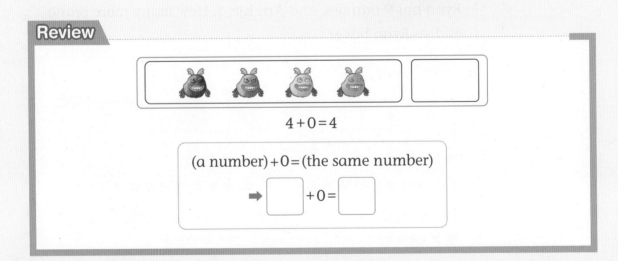

$4 + 0 = 4$

(a number) + 0 = (the same number)

➡ ☐ + 0 = ☐

Comics for Learning Math

When you add 0 to a number, the value does not change.

Review

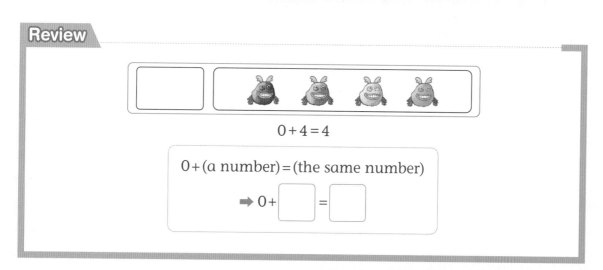

$0 + 4 = 4$

$0 + $ (a number) $=$ (the same number)

➡ $0 + \boxed{} = \boxed{}$

5 Basics of Addition and Subtraction

Adding and Subtracting 0

Comics for Learning Math

If you subtract a number from itself, you will get 0. If you subtract 0 from a number, the number stays the same.

Review

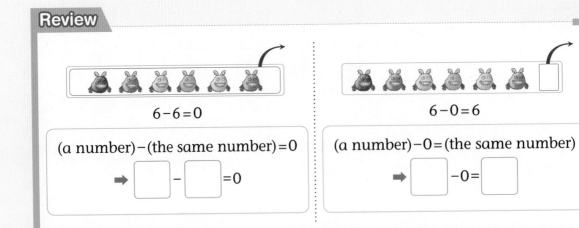

$6 - 6 = 0$

(a number) − (the same number) = 0

➡ ☐ − ☐ = 0

$6 - 0 = 6$

(a number) − 0 = (the same number)

➡ ☐ − 0 = ☐

Warm Up 05 A and B are the same number. Write that number on the line below.

There are 3 strawberries on one dish and none on the other. How many strawberries are there in total?

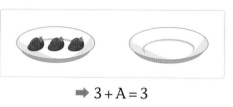

➡ $3 + A = 3$

There were 3 strawberries on the dish, but Ryan ate them all. How many strawberries are left?

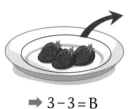

➡ $3 - 3 = B$

05-1 Which of the following equations has the greatest number value?

a. $8 + 0 = A$
b. $0 + 9 = B$
c. $7 - 7 = C$
d. $7 - 0 = D$

05-2 Which of the following is NOT correct?

a. If you add 0 to a number, the value does not change.
b. If you subtract 0 from a number, the value does not change.
c. $0 + 7 = 7$
d. $7 - 7$ cannot be calculated.

6. Basics of Addition and Subtraction

Understanding the Commutative Law of Addition

Comics for Learning Math

The order in which two numbers are added does not matter.
The total will be the same.

Adding 2 and 7 produces the same total as adding 7 and 2.

Review

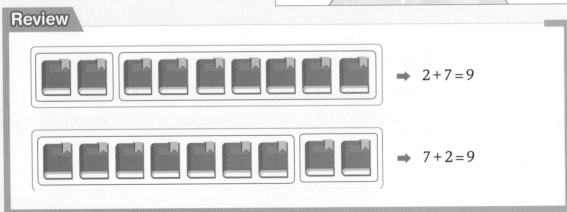

Warm Up 06 Look at the picture below and write the correct number in the box.

$$4 + 3 = \boxed{} + 4$$

06-1 Compare the two expressions. Write >, =, or < in the ◯.

(1) 3 + 5 ◯ 5 + 3 (2) 4 + 2 ◯ 2 + 4

06-2 Read the dialogue below. Choose who gave the correct explanation.

> Ryan: Aris, you got 7 stickers yesterday and 2 more today, right? I got 2 stickers yesterday and 7 more today, so I have more stickers than you do.
> Aris: No, that's not true. Yesterday, I got more stickers than you, so I still have more.
> Namute: You both have the same number of stickers because the total is the same no matter what order you add the numbers.

Basics of Addition and Subtraction

Number Pairs of 10 (1)

Comics for Learning Math

There are many different pairs of numbers that add up to 10.

The number pairs of 10 are 1 and 9, 2 and 8, 3 and 7, 4 and 6, and 5 and 5.

Review

The number pairs of 10 are:
➡ 1 and 9, 9 and 1, 2 and 8, 8 and 2, 3 and 7, 7 and 3, 4 and 6, 6 and 4, and 5 and 5.

Warm Up 07 How many more fans do we need to have 10 fans?

07-1 A bakery is selling corn bread and almond bread in sets of 10. Who made a mistake?

a. Ryan: I picked 2 corn and 8 almond.
b. Aris: I picked 5 corn and 5 almond.
c. Namute: I picked 6 corn and 3 almond.
d. Gilly: I picked 9 corn and 1 almond.

07-2 Look at the 3 pairs of number cards. Can you see a pattern? Which of the following explanations is <u>NOT</u> correct?

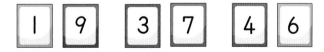

a. 2 cards of 5 would follow this same pattern.
b. When you add the numbers in each pair, you will get 10.
c. Even if you change the order of the cards and then add the numbers, the sum will be the same.
d. Instead of 1 and 9, you should have 2 and 9.

Basics of Addition and Subtraction

Number Pairs of 10 (2)

Comics for Learning Math

The number pairs that add up to 10 are 1 and 9, 2 and 8, 3 and 7, 4 and 6, and 5 and 5.

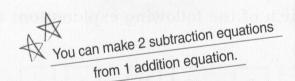

⭐⭐ You can make 2 subtraction equations from 1 addition equation.

Review

You can create subtraction equations by using number pairs of 10.

ex) $1+9=10$ ➡ $10-1=9, 10-9=1$ ex) $2+8=10$ ➡ $10-2=8, 10-8=2$

ex) $3+7=10$ ➡ $10-3=7, 10-7=3$ ex) $4+6=10$ ➡ $10-4=6, 10-6=4$

ex) $5+5=10$ ➡ $10-5=5$

Warm Up 08 You bought 10 strawberries and then ate 3. How many strawberries are left?

08-1 Find the equation that correctly matches the number line.

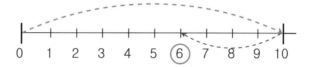

a. $6 + 4 = 10$ b. $4 + 6 = 10$
c. $10 - 4 = 6$ d. $10 - 6 = 4$

08-2 Find A and B. Then insert the numbers into the equation to find the answer.

- $10 - 2 = (\ A\ )$
- $10 - (\ B\ ) = 7$

➡ $A - B = \boxed{}$

Workbook 157

9 Basics of Addition and Subtraction

Adding (Tens) + (Ones)

Comics for Learning Math

The addition equation is
14 + 3 = 17.

★★ Adding 2 numbers is easier if you use number blocks.

Review

How to Calculate 14 + 3

1st Using Number Blocks

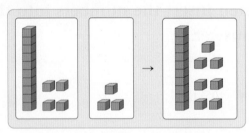

2nd Adding in Vertical Columns

```
  1 4            1 4            1 4
+   3      →   +   3      →   +   3
                   7            1 7
```

Warm Up 09 There are 12 apples and 4 oranges in the box. How many pieces of fruit are there altogether?

09-1 Choose the number line that correct expresses the addition expression 15 + 2.

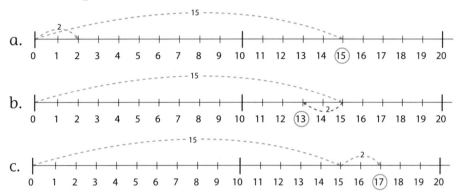

09-2 Here are three explanations for solving 13 + 6. Choose the explanation that is NOT correct.

> A. Separate 13 cubes into a group of 10 and a group of 3. Add 6 more cubes to the group of 3 for a total of 9 cubes. 10 plus 9 is 19.
>
> B. Move forward 16 spaces on a number line, and then move 3 spaces back. You will stop on 13, so 13 + 6 = 19.

10 Basics of Addition and Subtraction

Adding Three Numbers (The Associative Law)

> When you are adding three numbers, you can add them in any order.

Comics for Learning Math

Addition is easy if you know the number pairs of 10. Even with bigger numbers, you can find the answer quickly if you look for 10.

★ Adding three numbers is easy if you add two numbers first.

Review

Calculate $7 + 3 + 6$ (Find a number pair of 10).

1st Using Number Blocks

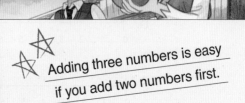

➡ $7 + 3 + 6 = 16$

2nd Using a Number Pair of 10

$7 + 3 + 6 = 16$
 └─10─┘
 └──16──┘

160 Ryan's Math Adventure 2

Warm Up 10 Solve the addition equation.

$$5 + 5 + 3 = \boxed{}$$

10-1 Choose the equation that has a number pair of 10 correctly underlined.

a. $\underline{2+8}+7=17$
b. $\underline{2}+8+\underline{7}=17$
c. $\underline{5+4}+6=15$
d. $1+\underline{3+9}=13$

10-2 Choose the sentence that is <u>NOT</u> correct about the addition expression $6+8+4$.

a. You can find the answer easily if you find a number pair of 10.
b. 6 plus 4 is 10.
c. $6+8+4$ is the same as 10 plus 4.
d. If you add all three numbers, the total is 18.

11 Basics of Addition and Subtraction

Calculating (Ones) + (Ones) = (Tens)

Comics for Learning Math

There are other ways to do addition.

1st

```
    7 + 5
   /   \
  2 + 5 + 5
      └───┘
    2 + 10
```

2nd

```
    7 + 5
     / \
  7 + 3 + 2
  └───┘
   10 + 2
```

Try to find number pairs of 10 first.

Review

Calculate 6 + 8

1st Using Number Blocks

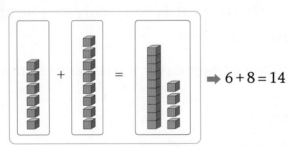

➡ 6 + 8 = 14

2nd Separating Numbers to Find Number Pairs of 10

```
    6 + 8              6 + 8
   /   \              /   \
  4 + 2 + 8         6 + 4 + 4
      └───┘         └───┘
    4 + 10           10 + 4
```

➡ 6 + 8 = 14

Ryan's Math Adventure 2

Warm Up 11 Write the number that goes into the box ☐.

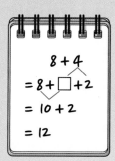

$$8 + 4$$
$$= 8 + \boxed{} + 2$$
$$= 10 + 2$$
$$= 12$$

11-1 Choose the correct A and B.

$$9 + 6$$
$$= 9 + \boxed{A} + \boxed{B}$$
$$= 10 + \boxed{}$$
$$= 15$$

a. A = 1, B = 5   b. A = 2, B = 4   c. A = 4, B = 2
d. A = 5, B = 0   e. A = 5, B = 1

11-2 Choose the person who correctly explained how to solve 7 + 6.

Ryan: If you separate 6 into 2 and 4, you find a number pair of 10.

Aris: If you separate 7 into two numbers and add 6, your answer will be different than if you separated 6 into two numbers and added 7.

Namute: Separate 7 into two numbers and add 3 + 4 + 6. Now you can find a number pair of 10.

Basics of Addition and Subtraction

Calculating (Tens) − (Ones)

Comics for Learning Math

10 students at the magic school have learned magic.
13 − 3 = 10

✩ You can change 1 stack of ten into 10 blocks.

Review

Calculate 13 − 3

1st Using Number Blocks

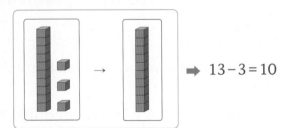

➡ 13 − 3 = 10

2nd Using a Number Line

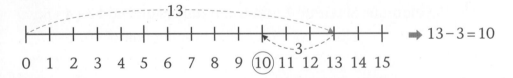

➡ 13 − 3 = 10

Warm Up 12 Look at the picture and write the correct number in the box.

 ➡ 18 − 5 = ☐

12-1 Look at the number line. Choose the subtraction equation that matches the line.

a. 14 − 2 = 11   b. 14 − 3 = 11   c. 15 − 3 = 11

12-2 Choose the answer choices that best matches the **Story**.

Story
There were 16 oranges. Ryan ate 4. How many oranges are left?

a.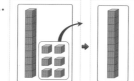

b.

c. 16 − 4 = 13

d.

13 Basics of Addition and Subtraction

Subtracting Three Numbers

Comics for Learning Math

Subtracting three numbers isn't difficult!

Subtract the numbers in order starting from left to right.

Review

Calculate $8 - 3 - 2$

1st Using Pictures

➡ $8 - 3 - 2 = 3$

2nd Using a Number Line

➡ $8 - 3 - 2 = 3$

Warm Up 13 Ryan took 4 oranges. Aris took 3 oranges. How many oranges are left?

13-1 Ryan had 16 coupons. Then he bought a robot and a book. How many coupons does he have left?

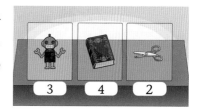

13-2 Choose the picture that best matches the Story.

> **Story**
> There were 12 pens. Aris took 2 pens, and Ryan took 5. How many pens are left?

a.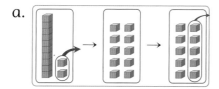

b.

c.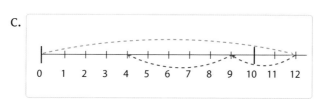

d. $12 - 2 - 5 = 3$

Basics of Addition and Subtraction

Adding and Subtracting Three Numbers

Comics for Learning Math

When an expression has both addition and subtraction, solve the numbers in order from left to right.

You must solve an expression like 9 − 3 + 1 in order. 9 − 3 = 6. 6 + 1 = 7. Don't switch the order!

Solving the numbers in order from left to right can make difficult expressions easy to solve.

Review

Calculate 5 + 4 − 3

1st Using Pictures

➡ 5 + 4 − 3 = 6

2nd Using a Number Line

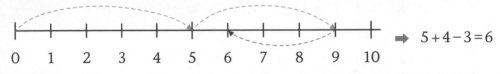

➡ 5 + 4 − 3 = 6

Warm Up 14

Look at the picture and write the correct number in the box.

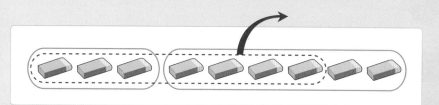

➡ 3 + 6 − 7 = ☐

14-1
Choose the answer choice that does NOT follow the correct steps to solve 8 − 3 + 4.

A. 8 − 3 + 4 = 9
     5
     9

B. 8 → 5
   − 3   + 4
     5   9

C. 8 − 3 + 4 = 1
         7
     1

14-2
Choose the answer that will make this equation true.

7 ◯ 1 ◯ 2 = 8

a. +, +
b. +, −
c. −, +
d. −, −

15. Basics of Addition and Subtraction

Calculating (Tens) − (Ones) = (Ones)

Comics for Learning Math

Let's look at other ways of solving expressions!

$$12 - 4$$
$$= 10 - 4 + 2$$
$$= 6 + 2$$
$$= 8$$

☆☆ Using number blocks can help make subtraction easier.

Review

Calculate 12 − 4

1st Using Number Blocks (1)

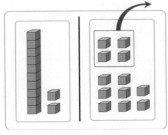

➡ 12 − 4 = 8

2nd Using Number Blocks (2)

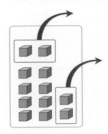

$$12 - 4$$
$$= 12 - 2 - 2$$
$$= 10 - 2$$
$$= 8$$

Warm Up 15 Look at the picture and write the correct number in the box.

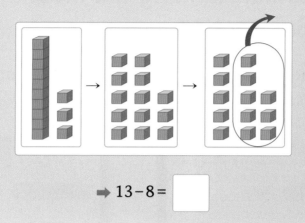

➡ 13 − 8 = ☐

15-1 Choose the equation that has a different answer from the others.

a. 12 − 5   b. 14 − 7   c. 11 − 4
d. 16 − 8   e. 15 − 8

15-2 Someone has made a mistake when solving 16 − 7. Was it Ryan or Aris?

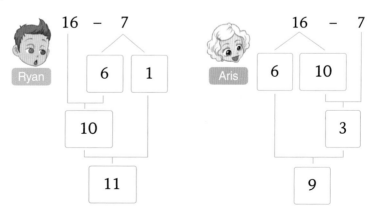

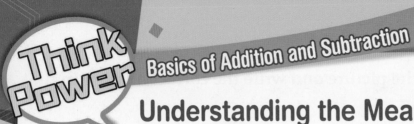

Understanding the Meaning of Subtraction

The following is from Ryan's diary. Read the diary and answer the questions.

1.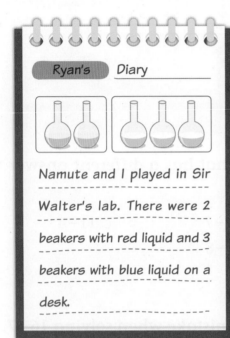

   Ryan's Diary

   Namute and I played in Sir Walter's lab. There were 2 beakers with red liquid and 3 beakers with blue liquid on a desk.

2.

   Ryan's Diary

   Namute shook the desk. 1 beaker fell off and broke.

3.

   Ryan's Diary

   Namute panicked. He took 2 beakers from a shelf and placed them on the desk.

4.

   Ryan's Desk

   I started to worry about how many beakers were now on the desk compared to before.

1. Read the diary again carefully and write down four equations from the story.

   1.
   2.
   3.
   4.

2. Draw a picture to match the 4th page of the diary.

Workbook 173

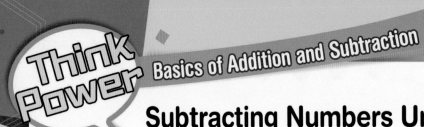

Basics of Addition and Subtraction

Subtracting Numbers Under 20

The following is a conversation between Ryan and Aris. What would be the answer for ▢? Write the steps to solve this question and the answer.

1.

2.

Think Power — Basics of Addition and Subtraction

Adding Numbers Under 20

Ryan and Aris have invited some friends over to their house. They are choosing among 3 potato dishes to make with Sir Walter. Which ones should they make? Choose 3 dishes and find the total number of potatoes they need to buy from the market.

Potato Salad

Ingredients
3 potatoes, 1 ear of corn, 1 bell pepper, some mayonnaise, a carrot, some sugar, some salt.

Potato Soup

Ingredients
2 potatoes, a young pumpkin, some kelp, some anchovy, some salt, some water, some flour.

Fried potato dumpling

Ingredients
8 potatoes, some whipping cream, some bread crumbs, some salt, some cooking oil.

French Fries

Ingredients
6 potatoes, some cooking oil.

3 dishes that you want to make:

(,   ,   )

Number of potatoes needed to buy from the market:

(   )

Basics of Addition and Subtraction

Subtracting Numbers Under 20

In music, you write musical notes on a staff to make a song. Look at the following song sheet and answer the questions.

Twinkle, Twinkle, Little Star

1. Sing "Twinkle, Twinkle, Little Star" with Do Re Mi Fa Sol La Ti Do. Which note repeats itself the most?

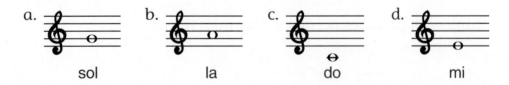

a. sol   b. la   c. do   d. mi

2. How many more times does Sol appear compared to Mi?

Answers

1 Understanding Addition (1) p. 141

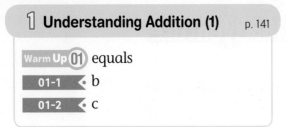

Warm Up 01

With 3 + 1 = 4, you say 3 plus 1 equals 4.

01-1

In order for 4 + 3 = 7 to work, the correct answer should be the jar with 3 beads.

01-2

According to the number line, you can say "7 plus 2 equals 9." or "Adding 7 and 2 equals 9."

2 Understanding Addition (2) p. 143

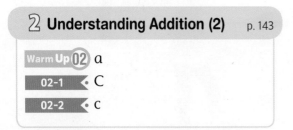

Warm Up 02

You need to add 3 blue balloons and 2 pink ones. The correct addition equation is 3 + 2.

02-1

The correct addition equation this picture represents is 6 + 3 = 9, not 6 + 2 = 8.

02-2

In C, there are 6 sparrows sitting in the trees. 2 of them are flying away. There are 8 sparrows in total.

3 Understanding Subtraction (1) p. 145

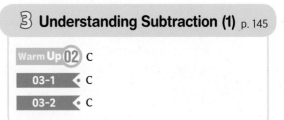

Warm Up 03

If you take away 3 from 5, like in the picture, you can find the number of remaining melons.

03-1

On the number line, each point represents a number. If you start at 0, move forward 8 points and backwards 3, you will stop at 5. The subtraction equation for this is 8 − 3 = 5.

03-2

Find the picture that represents the number of cookies remaining. To show this in an equation : 6 − 2 = 4

Answers

4 Understanding Subtraction (2) p. 147

Warm Up 04 2
04-1 9 − 3 = 6
04-2 d

Warm Up 04

Compare the number of bananas Ryan and Aris have in the picture. Ryan has 2 more bananas than Aris.

04-1

If you want to express how many more blue pencils there are compared to red pencils in a subtraction equation, you need to write the bigger number first and express it as 9 − 3 = 6.

04-2

Find the picture that represents 9 is 4 more than 5. If you write this as a subtraction equation, it is 9 − 5 = 4.

5 Adding and Subtracting 0 p. 151

Warm Up 05 0
05-1 b
05-2 d

Warm Up 05

If you add 0 to a number, the value does not change. Also, if you subtract the exact same number as the original number, the answer will be 0.

05-1

If you add 0 to any number, the value does not change. A = 8, B = 9, C = 0, and D = 7.

05-2

If you add 0 to or subtract 0 from a number, the value does not change. If you subtract a number from itself, you will get 0. d. 7 − 7 = 0

6 Understanding the Commutative Law of Addition p. 153

Warm Up 06 3
06-1 (1) =   (2) =
06-2 Namute

Warm Up 06

Even if you change the order of the numbers and then add them, the answer will be the same.

06-1

Even if you change the order of the numbers and then add them, the answer will be the same.

Answers

06-2
Even if you change the order of the numbers and then add them, the answer will be the same.

7 Number Pairs of 10 (1)   p. 155

Warm Up 07 6 fans
07-1 c
07-2 d

Warm Up 07

Let's count on our fingers. Spread open all 10 fingers and then fold down 4 fingers to represent the number of fans we already have. Count the rest of your fingers. Now you can see that 4 fans and 6 fans would make 10 fans in total.

07-1
If you add the number of loaves of corn bread and almond bread that Namute picked, it is 6 + 3 = 9. So the total is not 10. The total is 9.

07-2
d. If you add 2 and 9, it can't be 10.

8 Number Pairs of 10 (2)   p. 157

Warm Up 08 7 strawberries
08-1 c
08-2 5

Warm Up 08
If you eat 3 out of 10 strawberries, it is 10 − 3 = 7. There would be 7 strawberries left.

08-1
If you start at 10 and move back 4 spaces, you will be at 6. The subtraction equation is 10 − 4 = 6.

08-2
A equals 8, and B equals 3. A − B is 8 − 3, so the answer is 5.

9 Adding (Tens) + (Ones)   p. 159

Warm Up 09 16
09-1 C
09-2 B

Warm Up 09
There are 12 apples and 4 oranges in the box. The total number is 16.

4 Ryan's Math Adventure 2

Answers

09-1
To solve 15 + 2, move forward 15 spaces. Then move 2 more spaces.

09-2
To solve 13 + 6, move forward 13 spaces. Then move 6 more spaces.

10 Adding Three Numbers (The Associative Law)   p. 161

Warm Up 10 13
10-1 a
10-2 c

Warm Up 10
In the equation 5 + 5 + 3, 5 and 5 make 10. Adding 3 to 10 makes 13 in total.

10-1
In 2 + 8 + 7, adding 8 and 2 makes 10. Then add 7 to get 17.

10-2
6 + 8 + 4 is the same as 10 + 8.

11 Calculating (Ones) + (Ones) = (Tens)   p. 163

Warm Up 11 2
11-1 a
11-2 Namute

Warm Up 11
After splitting 4 into 2 and 2, add the 8 and 2 to make 10.

11-1
After separating 6 into 1 and 5, add 9 and 1 to make 10. Then add 10 and 5 to make 15.

11-2
Namute: When you split 7 into 3 and 4, you can find the number that can be added with 6 to make 10.

12 Calculating (Tens) − (Ones)   p. 165

Warm Up 12 13
12-1 b
12-2 d

Warm Up 12
When you count the remaining acorns, you'll find there are 13 left, so 18 − 5 = 13.

Workbook Answers 5

Answers

12-1
You move 14 spaces to the right and then move backward 3 spaces, which can be expressed as 14 − 3 = 11.

12-2
In the Story, the correct subtraction equation is 16 − 4 = 12.
a. 16 − 6 = 10, b. 16 − 3 = 13,
c. 16 − 4 = 12

13 Subtracting Three Numbers p. 167

Warm Up 13 2 oranges
13-1 9 coupons
13-2 a

Warm Up 13
Out of 9 oranges, 4 oranges to Ryan and 3 oranges to Aris leaves 2 oranges left.

13-1
(Original amount of coupons) − (Coupons needed to purchase a robot) − (Coupons needed to purchase book) = (Leftover number of coupons)
→ 16 − 3 − 4 = 9

13-2
If you express the Story as a subtraction equation, it will be 12 − 2 − 5 = 5
a. 12 − 2 − 5 = 5
b. 12 − 2 − 4 = 6
c. 12 − 3 − 5 = 4
d. 12 − 2 − 5 = 3 (×) → 12 − 2 − 5 = 5 (o)

14 Adding and Subtracting Three Numbers p. 169

Warm Up 14 2
14-1 C
14-2 c

Warm Up 14
As you can see in the picture, you have 2 erasers left, so 3 + 6 − 7 = 2.

14-1
C. When you have addition and subtraction in the same expression, you must solve from left to right.

14-2
a. 7 + 1 + 2 = 10   b. 7 + 1 − 2 = 6
c. 7 − 1 + 2 = 8   d. 7 − 1 − 2 = 4

15 Calculating (Tens) − (Ones) = (Ones) p. 171

Warm Up 15 5
15-1 d
15-2 Ryan

Warm Up 15
With 13 − 8, you need to change the tens block into ones and then subtract 8. → 13 − 8 = 5

15-1
a. 12 − 5 = 7 b. 14 − 7 = 7 c. 11 − 4 = 7
d. 16 − 8 = 8 e. 15 − 8 = 7

6 Ryan's Math Adventure 2

Answers

15-2

Ryan split 7 into two numbers. Aris split 16 into two numbers. Ryan took away 6 from 16 to make 10. He should take 1 from 10 to get the answer. The answer is 9.

Understanding the Meaning of Subtraction
p. 172~173

1 1. 2 + 3 = 5   2. 5 − 1 = 4
     3. 4 + 2 = 6   4. 6 − 5 = 1

2

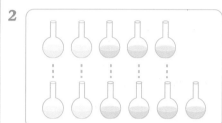

Subtracting Numbers Under 20
p. 174

1 (1) 3 + 2 = 5, 5 + 7 = 12
   or □ − 7 − 2 = 3, □ = 3 + 2 + 7

2 12

Adding Numbers Under 20
p. 175

The correct answer is the total number of potatoes you need for the 3 dishes you chose.

You need 3 potatoes for potato salad, 2 for potato soup, 3 for fried potato dumplings, and 6 for french fries. You can simply add the number of potatoes needed for the dishes you chose.

Answers

★ Let's check our work.
- Did you correctly add the number of potatoes needed for each dish?

Subtracting Numbers Under 20
p. 176

1 a
2 2 times

1 Do: 6 times, Mi: 8 times, La: 4 times, Sol: 10 times.
2 $10 - 8 = 2$, so Sol appears 2 more times than Mi.